ULTIMATE SUPERCARS

McLAREN 720S

By John Perritano

Kaleidoscope
Minneapolis, MN

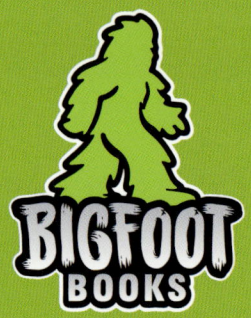

The Quest for Discovery Never Ends

This edition first published in 2021 by Kaleidoscope Publishing, Inc.

No part of this publication may be reproduced in whole or in part without written permission of the publisher.

For information regarding permission, write to
Kaleidoscope Publishing, Inc.
6012 Blue Circle Drive
Minnetonka, MN 55343

Library of Congress Control Number
2020936268

ISBN
978-1-64519-266-4 (library bound)
978-1-64519-334-0 (ebook)

Text copyright © 2021 by Kaleidoscope Publishing, Inc. All-Star Sports, Bigfoot Books, and associated logos are trademarks and/or registered trademarks of Kaleidoscope Publishing, Inc.

Printed in the United States of America.

FIND ME IF YOU CAN!

Bigfoot lurks within one of the images in this book. It's up to you to find him!

TABLE OF CONTENTS

Chapter 1: Like a Rocket Ship **4**

Chapter 2: Bruce McLaren's Car **12**

Chapter 3: A New Addition to the Super Series **18**

Chapter 4: Who's Faster? **24**

Beyond the Book *28*
Research Ninja *29*
Further Resources *30*
Glossary *31*
Index *32*
Photo Credits *32*
About the Author *32*

Chapter 1
Like a Rocket Ship

What do a spaceship and the McLaren 720S have in common? A lot! Ryan Conversano knows the answer, too. He's a rocket scientist at NASA. That means he knows a thing or two about space. He also knows about speed. The people at McLaren know speed, too. In fact, Conversano says the 720S is like a spaceship on wheels.

The 720S is fast. It can fly down the road at 212 miles per hour (341 kph).

Why is the car so speedy? Conversano says it has to do with how it moves through the air.

When the 720S is speeding down the road, air hits the front of vehicle. That generates **friction**. Friction slows a car down. McLaren drivers don't let that stop them, Conversano says. The 720S has an **aerodynamic** body. That means it cuts down on friction.

FUN FACT
The 720S's headlights are even shaped to make air flow more easily over the car.

One big reason is the splitter in front of the car. This car part forces air to flow above and below the vehicle. The splitter also creates downforce. Downforce allows the car to grip the road better.

The car's design makes it more stable. How do we know that? It can stop in 4.6 seconds from a speed of 124 miles per hour (200 kph)!

The 720S Spider is a convertible. It has panels on the roof that can be taken out. Drivers can feel the wind in their hair!

That's not all. There's a lot of force behind the 720S. The 720S power plant spits out 710 **horsepower**. That creates an incredible amount of force. Much of it comes from the car's electric **thrusters**, just like on some spaceships.

FUN FACT
The average American car only has about 200 horsepower.

PARTS OF A
MCLAREN 720S

Removable panels for Spider version

Engineers build spaceships as light as they can. The 720S's engineers do the same. The lighter the car, the faster the car. That's because it takes less force to move a light object. McLaren's engineers built much of the car using carbon. Carbon is strong and light.

FUN FACT
The 720S has a rare 7-speed transmission. Most cars don't have more than five.

V8 engine

Splitter

LED headlights

The 720S has a carbon "tub." The tub is part of its **chassis**. It's called the Monocage II. It surrounds the driver.

The 720S has a "layered" design. The car's body helps cut down on drag. Drag is what holds a moving car back. The 720S's body also helps the car

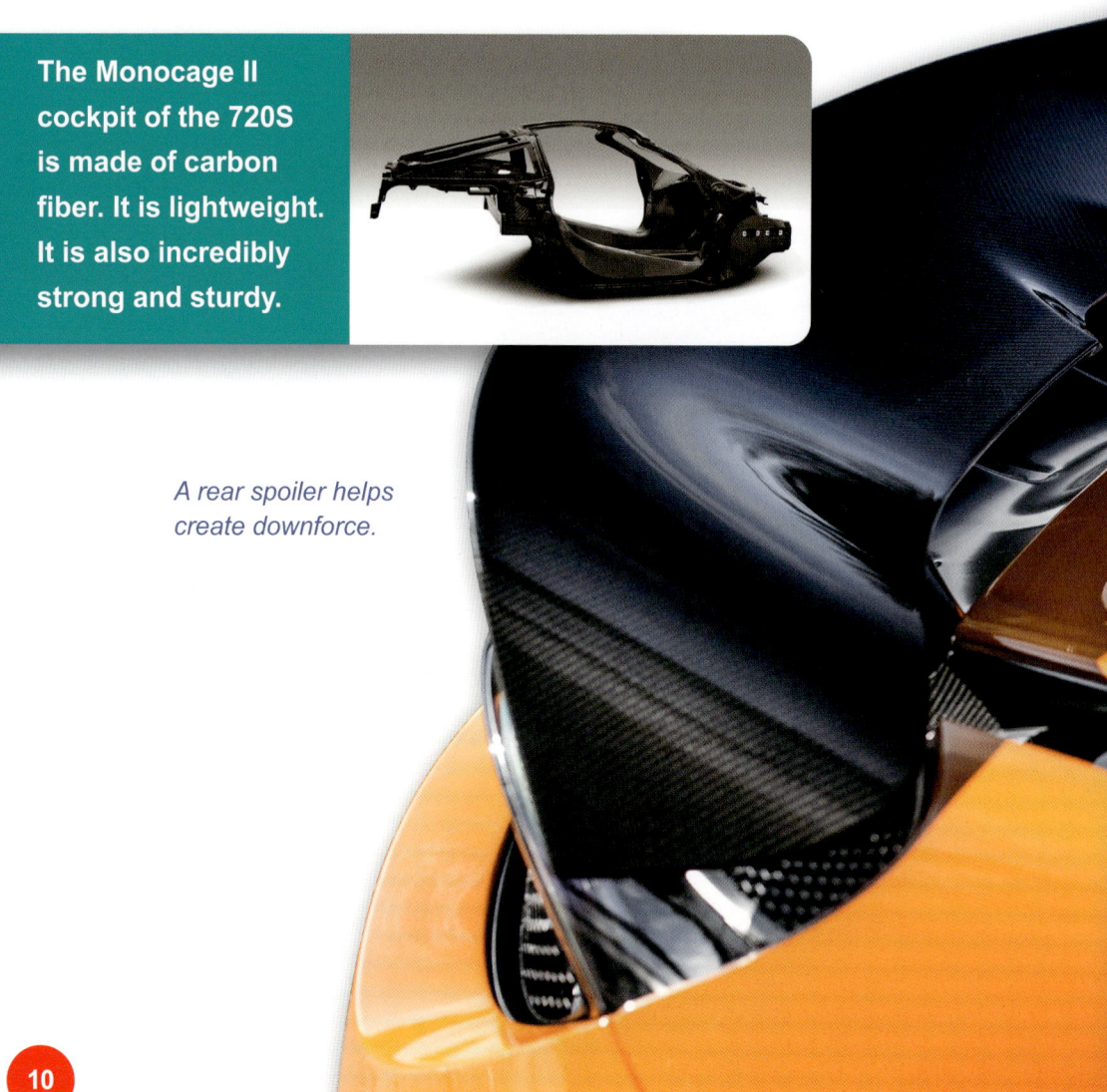

The Monocage II cockpit of the 720S is made of carbon fiber. It is lightweight. It is also incredibly strong and sturdy.

A rear spoiler helps create downforce.

mechanically. Parts of the body take moving air and use it to cool the engine.

All this science is the reason why the 720S is the spaceship of the supercars. Supercars are high performance vehicles that are **street legal**.

Chapter 2
Bruce McLaren's Car

Bruce McLaren was born in New Zealand in 1937. He spent lots of time hanging around his parents' garage. He loved motorcycle racing. Then he gave up motorcycles for race cars.

Bruce entered his first car race in 1952. He loved the thrill of speed. He became a pro race car driver and showed a lot of promise. He went to Europe to race with the best in the world.

McLaren drove for three racing teams. He won the 1959 United States Grand Prix. He was only 22. He'd win more races over the years. In 1965, he formed his own team.

An Austrian stamp featured McLaren

Bruce McLaren (left) after winning LeMans in 1966. Co-driver Chris Amon is at right.

McLaren drove this Ford car to victory.

McLaren was more than a driver. He was an engineer. He started building race cars in 1964. He won the 1968 Belgian Grand Prix in a car he built.

Sadly, McLaren died in 1970. He was testing one of his cars at a track in England. The car crashed. Before he died, McLaren had decided to stop racing. He wanted to spend more time designing race cars.

His racing team still survives. His amazing car company is still very busy, too.

WHERE THE MCLAREN 720S IS MADE

Woking, England: McLaren Technology Centre

FUN FACT
Since 1968, McLaren teams have won more than 180 F1 races.

For years McLaren cars ruled the race track. Today, its street cars compete with supercars from Ferrari, Lamborghini, and Porsche.

The McLaren F1 was the company's first road car. McLaren launched the car in 1992. At the time, it was the world's fastest. Its top speed was 242 miles an hour (389 kph).

The company introduced the 650S in 2014. It had a 3.8-liter twin-turbo V8 engine. It could go from 0 to 60 in just 3 seconds.

BUTTERFLY DOORS

One of the hallmarks of a McLaren car is the doors. Most car doors open outward on a hinge. McLaren doors have a hinge, too. But they raise up instead of opening out. Both doors tilt up to form a sort of L shape with the car's body. People sometimes call them butterfly doors.

One version of the 650S has a 666-horsepower motor. It can go from 0 to 60 in 2.9 seconds. McLaren would use the 650S as a step on the way to the 720S.

Chapter 3
A New Addition to the Super Series

The reporter looked across the floor. She and her photographer were at the 2017 Geneva Motor Show. Everyone was talking about one amazing car.

"There it is," she points. The photographer looks. Both stare at the McLaren 720S. "This isn't a car," the photographer says. "This is something out of the future."

The photographer was right. The 720S's glass roof looks like a jet fighter's. It's lighter than the 650S. It's faster, too.

"What makes this car so exciting?" the reporter asks a McLaren designer.

"The 720S shows how we've grown as a company," the designer says. "This is something brand new. We're so proud of it."

THE MCLAREN 720S IN DETAIL

Height: 3 feet, 11 inches (1.1 m)

Width: 7 feet, 1 inch (2.15 m)

LENGTH: 14 feet, 1 inch (4.5 m)

WEIGHT: 2,828 pounds (1,282 kg)

TOP SPEED: 212 miles per hour (341 kph)

TIME FROM 0-60 MPH: 2.9 seconds

COST: $299,000 (United States)

The 720S is part of McLaren's "Super Series." It is the first of 15 new models McLaren is planning for the future. The 720S was inspired by Bruce McLaren's first F1 race car. The 720S can be driven on a highway. But it's also fast enough for the track.

The design is brave and bold. Engineers had several things in mind when building the 720S: speed, stability, fun, and luxury.

The 720S cockpit is different than in other McLarens. The cabin is more driver-friendly.

The digital instrument panel is behind the steering wheel. The driver can see the car's vital information. The 411 includes a speedometer. It also displays the **tachometer**. The driver can also see what gear the car is in.

Engineers worked hard to make the 720S easy to use. A person can open the butterfly doors even

in tight places. There's even luggage space behind the seats.

The car's interior mixes high-tech with racing elements. The seats are leather. A touch screen sits at the center of the dash. It's angled toward the passenger. **Toggle switches** allow the driver to throw the car into drive. The switches are placed below the dash's touch screen.

FUN FACT
On the McLaren website, you can design your own custom 720S.

Chapter 4
Who's Faster?

Lamborghini makes a fast car. One model is the Aventador Roadster. It can go from 0 to 60 in 3 seconds. It can fly down the road at 217.48 miles per hour (350 kph). But can it beat a 720S Spider?

Both cars together generate 1,460 horses. But does the McLaren have what it takes? It's time to find out who will win a quarter-mile drag race.

Two drivers get in the cars. "Let's do this," the Lambo driver says.

Three . . . two . . . one! GO!

The cars roar down the track!

The Lamborghini surges ahead!

The McLaren driver is not worried. His car is catching up.

"Oh, no!" the Lamborghini driver thinks. He sees the 720S coming up on him.

Both drivers press down on the gas. Both motors whine. The sound is deafening.

Then the McLaren passes the Lambo.

"See, ya!"

What happened? The rear wheel-drive McLaren edged out the Lambo. The 720S went from 0 to 60 in 2.9 seconds. It's quarter-mile time was 10.4 seconds. The Lambo went from 0 to 60 in 3.0 seconds. It only lost the race by 0.1 second.

McLarens are all about speed. Next up? How about a McLaren spaceship?

FUN FACT
Why 0 to 60? That is the standard for comparing speed for all types of cars.

BEYOND
THE BOOK

After reading the book, it's time to think about what you learned. Try the following exercises to jumpstart your ideas.

RESEARCH

FIND OUT MORE. Where would you go to find out more about your favorite cars? Find out what company makes the car and locate its website. What information do the companies provide? What other sources of car information can you find?

CREATE

GET ARTISTIC. Cars start with creative artists and designers. Time for you to take a shot! Get art materials and create a great, new car. Will you make it a sports car? A sedan? A race car? What colors will you paint it? What features can you give it? Let your imagination go for a spin!

DISCOVER

DIG DEEPER. Bruce McLaren went from being a driver to a car designer. Inspired by his move, how would you change your family car to be more like a race car? Or come up with an all-new race car design. What features would it have? What kind of engine? How fast would it go?

GROW

GO TO A CAR SHOW. Car shows are a great way to see lots of cool cars up-close. Check your local events calendar, or ask at a car dealer for upcoming events. You can find shows of old cars and new cars, sports cars and classic cars. Go to a show and find a new favorite car to love!

RESEARCH NINJA

Visit www.ninjaresearcher.com/2664 to learn how to take your research skills and book report writing to the next level!

RESEARCH

DIGITAL LITERACY TOOLS

SEARCH LIKE A PRO
Learn about how to use search engines to find useful websites.

FACT OR FAKE?
Discover how you can tell a trusted website from an untrustworthy resource.

TEXT DETECTIVE
Explore how to zero in on the information you need most.

SHOW YOUR WORK
Research responsibly—learn how to cite sources.

WRITE

GET TO THE POINT
Learn how to express your main ideas.

PLAN OF ATTACK
Learn prewriting exercises and create an outline.

DOWNLOADABLE REPORT FORMS

Further Resources

BOOKS

Friedman, David. *McLaren Sports Racing Cars.* Minneapolis, MN: Motorbooks, 2000.

Garstecki, Julia. *McLaren 720S.* North Mankato, MN: Black Rabbit Books, 2019.

Mason, Paul. *British Supercars.* Mankato, MN: PowerKids Press, 2018.

WEBSITES

Factsurfer.com gives you a safe, fun way to find more information.

1. Go to www.factsurfer.com.
2. Enter "McLaren 720S" into the search box and click 🔍
3. Select your book cover to see a list of related websites.

Glossary

aerodynamics: science relating to the motion of gases and fluids flowing over a moving body.

chassis: frame of a motor vehicle.

friction: resistance one object encounters when moving over another.

horsepower: the measure of the power of an engine.

street legal: having all the equipment in the car that allows a person to drive it legally on the road.

tachometer: instrument that measures the speed of an engine usually in revolutions per minute.

toggle switch: an electronic switch that opens and closes an electrical circuit.

thrusters: small, directional engines.

V8: A V8 engine has eight cylinders in the shape of a V.

Index

Aventador Roadster, 24
Belgian Grand Prix, 14
body, 5, 10, 11, 16
Conversano, Ryan, 4, 5
engine, 8, 9, 11, 14, 16, 21, 22
England, 14, 15
Ferrari, 16
Geneva Motor Show, 18, 19
interior, 22, 23
Lamborghini, 16, 25, 26
McLaren, Bruce, 12, 13, 14, 21
McLaren 650S, 16, 17, 18
McLaren 720S Spider, 6, 8, 24
McLaren F1, 16
Monocage II, 10
NASA, 4
Porsche, 16
Super Series, 18, 21
transmission, 9
United States Grand Prix, 12

PHOTO CREDITS

The images in this book are reproduced through the courtesy of: AP Images: 12. Courtesy McLaren Media: 4, 6, 8, 10, 16, 20, 21, 22, 24, 26. Shutterstock: Spatuletail 12 stamp; E. Safronov 14; i. viewfinder 19.
Cover: Mesam/Shutterstock (car); YIUCHEUNG/Shutterstock (background, top); zhao jiankang/Shutterstock (background, bottom).

About the Author

John Perritano is an award-winning journalist, author, and editor from Southbury, Connecticut. He has authored numerous books and articles on subjects such as science, technology, history, and current events. He holds a master's degree in American History from Western Connecticut State University.